# Case Histories in
# **Human Physiology**
## Second Edition

**Donna Van Wynsberghe, Ph.D.**
University of Wisconsin-Milwaukee
*Department of Biological Sciences*
Milwaukee, Wisconsin

**Gregory M. Cooley, M.D.**
Medical College of Wisconsin
*Affiliated Hospitals*
Milwaukee, Wisconsin

**WCB** **Wm. C. Brown Publishers**

Dubuque, IA   Bogota   Boston   Buenos Aires   Caracas   Chicago
Guilford, CT   London   Madrid   Mexico City   Sydney   Toronto

**Book Team**

Editor *Colin H. Wheatley*
Developmental Editor *Kristine Noel*
Production Editor *Kay Driscoll*
Publishing Services Coordinator *Barbara Hodgson*

## Wm. C. Brown Publishers
A Division of Wm. C. Brown Communications, Inc.

Vice President and General Manager *Beverly Kolz*
Vice President, Publisher *Kevin Kane*
Vice President, Director of Sales and Marketing *Virginia S. Moffat*
Vice President, Director of Production *Colleen A. Yonda*
National Sales Manager *Douglas J. DiNardo*
Marketing Manager *Craig S. Marty*
Advertising Manager *Janelle Keeffer*
Production Editorial Manager *Renée Menne*
Publishing Services Manager *Karen J. Slaght*
Royalty/Permissions Manager *Connie Allendorf*

## Wm. C. Brown Communications, Inc.

President and Chief Executive Officer *G. Franklin Lewis*
Senior Vice President, Operations *James H. Higby*
Corporate Senior Vice President, President of WCB Manufacturing *Roger Meyer*
Corporate Senior Vice President and Chief Financial Officer *Robert Chesterman*

Cover and Interior Design by Jeanne Calabrese

A Times Mirror Company

Library of Congress Catalog Card Number: 94–78940

ISBN 0–697–13791–0

Printed in the United States of America by Wm. C. Brown Communications, Inc., 2460 Kerper Boulevard, Dubuque, IA 52001

10  9  8  7  6  5  4  3  2

# Contents

# Respiratory Case Histories  *57*

# Gastrointestinal Case Histories  *71*

# Renal Case Histories  *85*

# Endocrine Case Histories  *99*

# Reproductive Case Histories  *121*

# Acknowledgments

I would like to thank the nearly 15,000 students over the past 20 years who have helped generate in my mind and heart this compilation of case histories in medical physiology. I would also like to thank the following special people for their valuable suggestions, support and time during the various stages of this production: Marshall Dunning, Jeanne Seagard, Irene O'Shaughnessy, Reinhold Hutz, John Buntin, Pat Kattar-Cooley, and Mary Jane Werderman. The following list of reviewers deserve gratitude as well for their valuable input: David S. Bruce, *Wheaton College;* Terry L. Derting, *Murray State University;* Nancy E. Harris, *Elon College;* Herbert W. House, *Elon College.*

May these pages serve to strengthen the integrative and problem-solving skills of those who diligently use it.

*Donna Van Wynsberghe*

# How to Use
# Case Histories in Human Physiology

This compilation of case histories in physiology is designed to provide opportunities for integrative thinking and problem solving for undergraduate students in anatomy and physiology or physiology courses. These case histories can be used during lectures or as a part of laboratory or discussion sections. Each section of this text is best used after each corresponding unit is completed in your course in anatomy and physiology or physiology.

These case histories are designed to teach you how to think and solve problems, both very rewarding experiences. Follow these guidelines:

1.  Read the appropriate case history and the accompanying questions.

2.  Define all appropriate terms.

3.  Compare all recorded values to the normal values and note whether the recorded values are greater or less than normal.

4.  Review the physiologic facts in your text pertaining to this case history.

5.  Use a medical dictionary or the dictionary in your text to clarify words or concepts you may not be familiar with.

6.  Use the reference laboratory values in Appendix B of this text if you are unsure of the normal clinical values for various parameters.

7.  Apply these physiologic facts to make your "diagnosis" and to answer the accompanying questions.

8.  Realize that you may have to *think for some time* before being able to put all the pieces together. Some of the answers are *not immediately obvious* but they are within your grasp.

9.  Have a good time doing this!

Answer Key to accompany *Case Histories in Human Physiology* is also available.

# Neurophysiology Case Histories

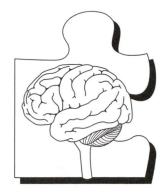

## C A S E  1

A 46-year-old female complained of weakness on the entire left half of her face, which quickly led to facial paralysis on that side. She was unable to wrinkle her forehead, smile, show her teeth, or puff out her cheek on the affected side. She was also unable to close her left eye. A temporary patch protected the exposed eye. Taste perception was distorted over the left anterior portion of the tongue. There was complete recovery of the facial paralysis and taste distortion within two months.

Name _____

Laboratory _____

Grade _____

# Neurophysiology Case Histories

## C A S E  1

**QUESTIONS:**

1. The cranial nerve involved in this individual is the _____ nerve.

2. This condition is known as _____ .

3. Why was there facial paralysis on the left half of her face; why was she unable to smile; why did she need a patch over her left eye?

4. Why was her taste distorted?

5. Compare *this* disorder to the major disorder associated with cranial nerve V.

6. Describe the normal functions of both of the cranial nerves discussed on previous page.

# Neurophysiology Case Histories

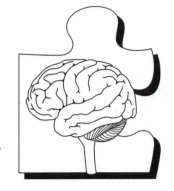

## CASE 2

A 30-year-old female has had several episodes, each separated by several months, in which she experiences brief periods of physical weakness, clumsiness of the legs and hands, visual disturbances, and mood swings. Her weakness is most prominent in her lower extremities. CT scans show evidence of multiple lesions in the white matter of the brain and spinal cord. *This is the clue for MS* During her last episode, there were increasing neurological deficits due to an increasing number of disseminated lesions. Prednisone (60 mg/day for five to seven days) was prescribed and helped her return to as normal and active a life as possible.

# Neurophysiology Case Histories

## C A S E  2

**QUESTIONS:**

1. What is the demyelinating disease affecting this individual?

2. What is myelin? What is its function?

3. How does the prednisone help this individual?

4. How are nerve cells ensheathed with myelin in the CNS? In the PNS?

# Neurophysiology Case Histories

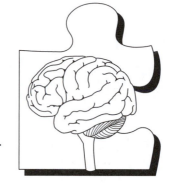

## C A S E   3

A 42-year-old male complained of severe, lower (lumbosacral) back pain which radiated to his left buttock, leg, and foot. This dull pain was intensified during coughing, sneezing, and bending. He also experienced muscle spasms in the left leg. There was mild sensory and motor loss in the left buttock and leg as well.

# Neurophysiology Case Histories

## C A S E  3

**QUESTIONS:**

1. What is the disorder of this individual?

2. What major nerve is involved? From what spinal cord levels does this nerve emerge?

3. Define dermatome and describe the dermatomes involved for this individual.

4. Why did coughing, sneezing, and bending intensify the dull pain?

5. Why was there both mild sensory *and* motor loss in the left buttock and leg?

6. What treatments may be of benefit to this individual?

## FOR YOUR CONSIDERATION: NEUROLOGY

How familiar are you with some common categories of neural drugs or agents? Match the drug category on the left with the indication or disorder on the right.

| | | |
|---|---|---|
| 1. | Anticonvulsants | _____ Cerebrovascular insufficiency |
| 2. | Cholinergics | _____ Mild, moderate, severe pain |
| 3. | Narcotic and opioid analgesics | _____ Myasthenia gravis |
| 4. | Vasodilators | _____ Seizures |

# Muscle Physiology Case Histories

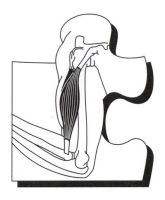

## C A S E  4

Parents of a 3-year-old boy noticed that their son was walking "on his toes," had a waddling gait, fell frequently and had difficulty getting up again, and was not able to run because of the difficulty in raising his knees. At age five, there was progressive muscular weakness and muscle wasting. Weakness of the trunk muscles led to increased lordosis and a protuberant abdomen. At age nine, he was confined to a wheelchair. Contractures appeared, first in the feet, as the gastrocnemius muscles tightened.

# Muscle Physiology Case Histories

## CASE 4

### QUESTIONS:

1.  This hereditary X-linked recessive disease characterized by progressive muscular weakness is

    _____ .

2.  What does dystrophy mean? Why is this term used to describe this case?

3.  What muscles would be involved in walking "on the toes"? Which muscles are "weakening"?

4.  Name the trunk muscles that weaken in certain cases of lordosis and abdominal protuberance.

# Muscle Physiology Case Histories

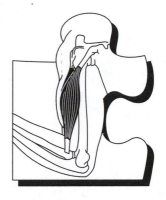

## CASE 5

A 17-year-old male was working vigorously on a summer construction crew building a new section of a freeway. In the intense heat of the day, he began to experience severe pain in the muscles of his limbs and carpopedal spasms. The cramping made his muscles feel like hard knots. The foreman of the crew instructed the young man to drink some salt water he had available and rest a while.

# Muscle Physiology Case Histories

## C A S E  5

**QUESTIONS:**

1. What is the cause of the cramping?

2. Describe carpopedal spasms.

3. Why is the ingestion of salt and water beneficial?

# Muscle Physiology Case Histories

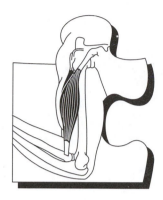

## C A S E  6

Prior to intubation for a surgical procedure, the anesthesiologist administered a single dose of the neuromuscular blocking agent, succinylcholine, to a 23-year-old male to provide muscular relaxation during surgery and to facilitate the insertion of the endotracheal tube. Following this, the inhalation anesthetic was administered and the surgical procedure completed.

# Muscle Physiology Case Histories

## CASE 6

### QUESTIONS:

1.  Beginning with depolarization at the neuromuscular junction, describe the normal sequence of events which lead to muscle contraction.

2.  What prevents acetylcholine (ACh) from accumulating in the neuromuscular junction and causing a sustained contraction in a normal individual?

3.  Succinylcholine acts as a depolarizing agent that prevents repolarization of the nerve. Therefore, no further ACh is released until the drug is cleared. Name another site within the neuromuscular junction that might be affected to prevent muscle contraction. (Hint: curare acts by this mechanism.)

# FOR YOUR CONSIDERATION: MUSCLE

How familiar are you with some common categories of musculoskeletal drugs or agents? Match the drug category on the left with the indication or disorder on the right.

1. Corticosteroids

2. Gold compounds

3. Nonsteroidal anti-inflammatory drugs (NSAIDS)

4. Skeletal muscle relaxants

_____ Muscle spasticity

_____ Osteoarthritis; bursitis

_____ Rheumatoid arthritis (RA)

_____ Severe inflammation

# Hematology Case Histories

## CASE 7

A 36-year-old male complained of fatigue, anorexia, weight loss, dyspnea on exertion, and a sense of abdominal fullness. An examination of peripheral blood and bone marrow resulted in the following:

**Complete blood count (CBC):**

| | |
|---|---|
| Erythrocytes | 2.8 million/cu mm |
| Leukocytes | 180,000/cu mm |
| Thrombocytes (platelets) | 53,000/cu mm |
| Hematocrit | 26% |

**Differential:**

Lymphocytes = 85% of all leukocytes

Bone marrow examination demonstrated a tremendous predominance of small lymphocytes (>90%). The bone marrow was so tightly packed that it aspirated with difficulty. The spleen was moderately enlarged, and there was generalized lymph node enlargement.

# Hematology Case Histories

## C A S E  7

**QUESTIONS:**

1. What is the primary disorder of this person?

2. Define leukocytosis.

3. What is the cause of the leukocytosis?

4. Define thrombocytopenia.

5. What is the cause of the thrombocytopenia?

6. Define anemia.

7. What is the cause of the anemia?

8. What is the cause of the feeling of abdominal fullness?

# Hematology Case Histories

## C A S E  8

A 14-year-old girl complained of fatigue and loss of stamina. Her appetite was marginal, as she was very conscious of maintaining her body weight at 96 pounds. Her monthly menstrual flow was always heavy and long from its onset at twelve years of age. Relevant laboratory findings included the following:

| | |
|---|---|
| Hematocrit (Hct) | 28% |
| Hemoglobin (Hgb) | 9 gm/dL |
| Iron | 16 ug/dL |
| Bone marrow iron | Absent |
| Erythrocytes | Small and pale |

Suggested treatment included ferrous sulfate or ferrous gluconate for six months orally between meals, since food may reduce absorption. A well-balanced diet was also suggested, as well as a gynecologic examination.

# Hematology Case Histories

## C A S E   8

**QUESTIONS:**

1.  What is the primary disorder of this individual?

2.  What does the ferrous sulfate or ferrous gluconate provide? Why is it necessary?

3.  What dietary inclusions would you suggest?

4.  Why is the gynecologic examination important?

5.  Why is bone marrow iron an important clinical indicator in this individual?

# Hematology Case Histories

## CASE 9

A 62-year-old male complained of weakness, headache, light-headedness, and fatigue. Upon physical examination, the following information was available:

| | |
|---|---|
| Erythrocytes | 8.5 million/cu mm |
| Leukocytes | 12,500/cu mm |
| Thrombocytes | 400,000/cu mm |
| Hct | 58% |
| $O_2$ saturation (arterial) | 94% |
| Serum erythropoietin | Undetectable |

The erythrocytes and leukocytes were immature in the peripheral blood smear. The spleen was enlarged. The determined therapy of choice was phlebotomy, 300–500 ml every other day, until the hematocrit was < 45%, with the possibility of myelosuppressive therapy, if needed.

# Hematology Case Histories

## C A S E  9

### QUESTIONS:

1. What is the disorder of this individual?

2. Why are the arterial $O_2$ saturation and erythropoietin levels important in making this decision?

3. Define phlebotomy.

4. How does phlebotomy help correct this problem?

5. Define myelosuppressive therapy.




6. Why may myelosuppressive therapy be needed?

# FOR YOUR CONSIDERATION: HEMATOLOGY

How familiar are you with some common categories of hematology drugs or agents? Match the drug category on the left with the indication or disorder on the right.

| | | | |
|---|---|---|---|
| 1. | Anticoagulants | _____ | Emboli; deep vein thrombosis |
| 2. | Hematinics | _____ | Hemophilia |
| 3. | Hemostatics | _____ | Heparin overdose |
| 4. | Heparin antagonists | _____ | Iron deficiency anemia |
| 5. | Vitamins ($B_{12}$) | _____ | Pernicious anemia |

# Cardiovascular Case Histories

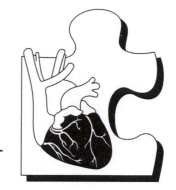

## C A S E  10

A 50-year-old airline pilot complained of severe, intense, precordial, crushing sensation with pain radiating to the left shoulder and down the inside of the left arm, triggered by an "off-duty" tennis match. The chest discomfort brought on by the exertion was relieved by rest. Emergency room examination resulted in the following information:

| | |
|---|---|
| Heart rate (HR) | 98 b/min |
| Blood pressure (BP) | 160/110 mm Hg |
| ECG: ventricular extrasystole arrhythmia (premature ventricular contraction [PVC]) as well as S-T segment depression and decreased R wave height | |

The following day an exercise tolerance test was performed to test the functional response to graded stress. This symptom-limited test gave an "ischemic" ECG response during exercise, characterized by a downward-sloping S-T segment. Mild exertion resulted in chest pain, which was relieved by sublingual nitroglycerin. Coronary angiography showed lumenal obstruction >70% (88%) in three major coronary vessels, including the left anterior interventricular (descending) coronary artery. Nitroglycerin, beta-blockers, and calcium channel blockers were tried as pharmacologic therapy. Angioplasty, the procedure in which a balloon-tipped catheter is inserted into the partially obstructed vessels, was able to increase coronary flow to near normal values.

# Cardiovascular Case Histories

## C A S E  10

### QUESTIONS:

1. What is the term for the chest pain experienced by this individual? What is the cause of this pain?

2. Draw a representative tracing of the Lead II ECG tracing as described in this case study for this individual and compare it to a normal tracing.

3. What is the site of action for each of the pharmacologic drug therapies?

   Nitroglycerin:

   Beta-blockers:

   Calcium channel blockers:

4. Describe the gross anatomy of normal coronary arteries.

5. Describe angioplasty. State the pros and cons of its use.

# Cardiovascular Case Histories

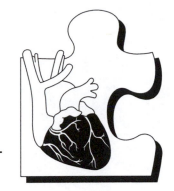

## C A S E  11

A 32-year-old nurse who had rheumatic fever as a child noticed a persistent tachycardia and light-headedness. Upon examination, chest x-rays showed an enlarged left atrium and left ventricle. ECG analysis showed atrial fibrillation. There was also mild pulmonary congestion. Cardiac evaluation resulted in the following information:

| | |
|---|---|
| Cardiac output (CO) | 3.4 L/min |
| Blood pressure (BP) | 100/58 mm Hg |
| Left atrial pressure (LAP) | 16 mm Hg |
| Right ventricular pressure (RVP) | 44/8 mm Hg |

Heart sounds revealed valvular regurgitation.

# Cardiovascular Case Histories

## CASE 11

**QUESTIONS:**

1.  Based on the information provided, which A-V valve is incompetent, allowing the regurgitation?

2.  Which heart sound would be pronounced and lengthened?

3.  Describe, using surface anatomy, the location at which this valvular disorder could best be heard.

4.  If the other A-V valve were incompetent instead of this one, would the CO, BP, LAP, and RVP be different? If so, how?

5. What are the causes of the tachycardia, light-headedness, and mild pulmonary congestion?

6. Calculate the pulse pressure (PP) and mean arterial blood pressure (MABP) for this individual.

   PP = _____

   MABP = _____

# Cardiovascular Case Histories

## C A S E   12

A 5'6", 210 lb, 64-year-old male business executive had a physical exam prior to his retirement from corporate work. His blood pressure was >180/115 on three separate days. Further examination showed normal to low plasma renin activity, elevated total peripheral resistance (TPR), cardiac output (CO) of 7.2 L/min, x-ray evidence of left ventricular hypertrophy, retinal hemorrhages, and mild polyuria. Recommended therapy was weight reduction to his ideal level, a low-salt diet (<2 gm/day sodium), prudent exercise, and a reduction in alcohol consumption (<3 oz whiskey/day). This change in life-style did little to change the condition. Medication was initiated in the form of an oral diuretic and progressed to a beta-blocker; eventually a vasodilator was included to reduce the blood pressure to <140/90.

# Cardiovascular Case Histories

## C A S E  12

### QUESTIONS:

1. What is the diagnosis for this individual?

2. What should this individual's ideal body weight be? (Assume he is of medium body frame.)

3. Explain the sites of action for the three pharmacologic agents prescribed for this individual.

   Oral diuretic:

   Beta-blocker:

   Vasodilator:

4. Why was the vasodilator not used before the other drugs were added?

5.  Outline the usual steps in the medical treatment of hypertension.

6.  What was the cause of the retinal hemorrhages?

7.  Define and indicate the causes of the cardiac hypertrophy and polyuria.

8.  Write the formulas for determining mean arterial blood pressure, blood pressure, and pulse pressure.

    MABP = _____

    BP = _____

    PP = _____

# FOR YOUR CONSIDERATION: CARDIOVASCULAR

How familiar are you with some common categories of cardiovascular drugs or agents? Match the drug category on the left with the indication or disorder on the right.

| | |
|---|---|
| 1. Adrenergics | _____ Essential hypertension |
| 2. Antianginals | _____ Pulmonary edema |
| 3. Antiarrhythmics | _____ Angina pectoris |
| 4. Antihypertensives | _____ Premature ventricular contractions; ventricular arrhythmias |
| 5. Antilipemics | _____ Hypotension |
| 6. Diuretics | _____ Reduction of low-density lipoprotein and total cholesterol levels in primary hypercholesterolemia |
| 7. Thrombolytic Enzymes | _____ Lysis of coronary artery thrombi after an acute myocardial infarction |

# Respiratory Case Histories

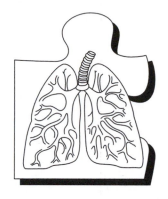

## CASE 13

A 17-year-old student has experienced reversible, periodic attacks of chest tightness with coughing, wheezing, and hyperpnea. She states that expiration is more difficult than inspiration. She is most comfortable sitting forward with arms leaning on some support. X-rays revealed mild overinflation of the chest. Results from laboratory and pulmonary function tests are as follows:

| | |
|---|---|
| Frequency | 20 breaths/min |
| Vital capacity (VC) | 2.9 L |
| $FEV_{1.0}$ | 1.4 L |
| $FEV_{1.0}/FVC$ | 56% |
| Functional residual capacity (FRC) | 3.89 L |
| Total lung capacity (TLC) | 6.82 L |
| $PaO_2$ | 70 mm Hg |
| $PaCO_2$ | 26 mm Hg |
| Pulse | 108 b/min |
| BP | 120/76 mm Hg |

Intermittent use of a bronchial smooth muscle dilator (1:1000 epinephrine by nebulizer) for several days caused marked improvement, resulting in the following laboratory and pulmonary function tests:

| | |
|---|---|
| VC | 4.15 L |
| $FEV_{1.0}$ | 3.1 L |
| $FEV_{1.0}/FVC$ | >75% |
| FRC | 3.7 L |
| TLC | 5.96 L |
| $PaO_2$ | 89 mm Hg |
| $PaCO_2$ | 38 mm Hg |
| Pulse | 129 b/min |
| BP | 122/78 mm Hg |

# Respiratory Case Histories

## C A S E 13

### QUESTIONS:

1.  What is the disorder of this 17-year-old student?

2.  Is this primarily a restrictive or an obstructive disorder? Why?

3.  Write the formula for determining residual volume (RV).

4.  Determine the residual volume (RV) before and after the use of the bronchodilator.

    RV before using the bronchodilator: _____

    RV after using the bronchodilator: _____

5.  Why is expiration more difficult than inspiration in this person?

6.  What does the change in pulmonary function after the bronchodilator therapy indicate?

7.  Why does the bronchodilator exaggerate the tachycardia?

8.  What causes the hypoxemia and the hypocapnemia in this person?

9.  A beta$_2$-adrenergic agent was prescribed for further use because it has less cardiostimulatory (beta$_1$) effect. Based on your knowledge of beta$_1$ and beta$_2$ receptors, why is this a good suggestion?

10. An anticholinergic agent was also suggested as a possible nebulizer agent. How might this help the breathing problem?

# Respiratory Case Histories

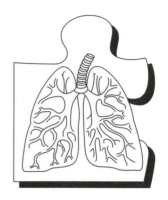

## C A S E  14

A 150 lb, 62-year-old man had a chronic productive cough, exertional dyspnea, mild cyanosis, and marked slowing of forced expiration. His pulmonary function and laboratory tests follow:

| | |
|---|---|
| Frequency | 16 breaths/min |
| Alveolar ventilation | 4.2 L/min |
| Vital capacity (VC) | 2.2 L |
| Functional residual capacity (FRC) | 4.0 L |
| Total lung capacity (TLC) | 5.2 L |
| Maximum inspiratory flow rate | 250 L/min |
| Maximum expiratory flow rate | 20 L/min |
| $PaO_2$ | 62 mm Hg |
| $PaCO_2$ | 39 mm Hg |

Pulmonary function tests after bronchodilator therapy:

| | |
|---|---|
| Frequency | 16 breaths/min |
| Alveolar ventilation | 4.35 L/min |
| VC | 2.4 L |
| FRC | 4.0 L |
| TLC | 5.2 L |
| Maximum inspiratory flow rate | 250 L/min |
| Maximum expiratory flow rate | 23 L/min |
| $PaO_2$ | 62 mm Hg |
| $PaCO_2$ | 38 mm Hg |

# Respiratory Case Histories

## C A S E  14

### QUESTIONS:

1. What is the disorder of this 62-year-old man?

2. Is this primarily a restrictive or an obstructive disorder? Why?

3. Why is the bronchodilator therapy ineffective in this man?

4. What causes the hypoxemia?

5. Calculate the residual volume (RV) for this person before and after the bronchodilator therapy. _____

6. What is the cause of this altered RV?

7. Calculate the tidal volume (TV) for this person before and after the bronchodilator therapy. _____

8. Is each TV normal or altered?

9. Calculate the minute ventilation (MV) for this person before and after the bronchodilator therapy. _____

10. Is each MV normal or altered?

# Respiratory Case Histories

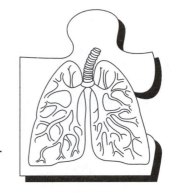

## CASE 15

A 22-year-old man was in a motorcycle accident with resultant neck injuries that led to partial paralysis of the upper and lower limbs. Almost immediately his chest felt heavy and he became dyspneic. His pulmonary function values were as follows:

| | |
|---|---|
| Vital capacity (supine) | 650 ml |
| Minute ventilation (supine) | 6 L/min |
| Respiratory rate (supine) | 30 b/min |
| $PaO_2$ | 61 mm Hg |
| $PaCO_2$ | 47 mm Hg |

# Respiratory Case Histories

## C A S E   15

**QUESTIONS:**

1.  What are the values for the tidal volume (TV) and the alveolar ventilation (AV) for this individual? (Assume a normal value for dead space.) Compare with normal.

    Tidal volume = _____

    Alveolar ventilation = _____

2.  What is contributing to the decreased alveolar ventilation?

3.  Which nerves and muscles are involved in this problem?

4.  How is the residual volume affected?

5.  Define dyspnea.

6.  What is causing the dyspnea in this individual?

7.  Define tachypnea.

8.  Describe the reflexes involved in causing the tachypnea in this individual.

## FOR YOUR CONSIDERATION: RESPIRATION

How familiar are you with some common categories of respiratory drugs or agents? Match the drug category on the left with the indication or disorder on the right.

1.  Adrenergics

_____ Coughs

2.  Antihistamines

_____ Rhinitis, allergy symptoms

3.  Antitussives and expectorants

_____ Bronchospasms, bronchial asthma

4.  Corticosteroids

_____ Severe inflammation or immunosuppression

# Gastrointestinal Case Histories

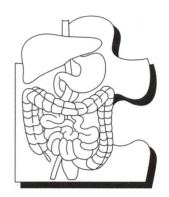

## C A S E   16

A 26-year-old business executive complained of a dull pain (heartburn) behind the sternum. The pain was postprandial (occurred after meals) and disappeared within a few minutes to an hour. It was often associated with belching and often was worse on lying down or on exertion after heavy meals. Sometimes it radiated to the back, jaws, shoulders, and down the inner aspects of the arms, simulating angina pectoris. X-rays revealed a small portion of the stomach above the diaphragm, and an endoscopic biopsy revealed mucosal inflammation. Esophageal manometry (determining pressures at the lower esophageal sphincter, LES) revealed decreased LES pressure. Esophageal pH monitoring showed reflux of gastric contents into the esophagus and provided direct evidence of gastroesophageal reflux. Recommended treatment for this individual is avoidance of strong stimulants of gastric acid secretion (e.g., coffee, alcohol) and avoidance of certain drugs (e.g., anticholinergics), and specific foods (fats, chocolates, whole milk, and orange juice), and smoking, all of which reduce LES competence. Elevation of the head of the bed by about six inches is also recommended. Suggested treatments also include the use of cholinergic agonists (e.g., bethanechol) and the use of histamine ($H_2$) antagonists (cimetidine). _hernia?_ _hytales_

# Gastrointestinal Case Histories

## C A S E   16

### QUESTIONS:

1.  What is the disorder of this 26-year-old business executive? Explain.

    *hiatal*

2.  What mechanisms normally prevent gastric reflux into the esophagus when lying down or bending over?

    *sphincters
    (LES) Lower ESP.*

3.  Why are anticholinergic agents avoided and cholinergic agonists recommended in the treatment of gastroesophageal reflux?

    *- Blocked*

    *antichol. blocks receptors
    ↓ ACh
    does to LES
    therefore the muscle would not close so it
    would ↓*

    *Cholinergic stimulates constrict and
    sphinter will shut*

4.  Why are histamine ($H_2$) antagonists recommended? *Block's receptors*

    *you need to block the histamine receptors*
    *that's why you need the antagonists*

5.  Why is elevation of the head of the bed recommended?

    *Because the gravity will help*

6.  What is the normal pH of the esophagus? Of the stomach? Predict values for the gastroesophageal patient in this case for lower esophageal and stomach pH.   *2-3*

    *Stomach will stay the same*
    *the lower esphageal 3-5*

# Gastrointestinal Case Histories

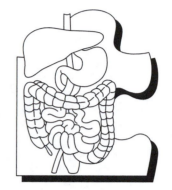

## C A S E   17

A 45-year-old store manager complained of a burning, gnawing pain, moderately severe, almost always in the epigastric region. The pain is absent when he awakens, appears in midmorning, and is relieved by food but recurs two to three hours after a meal. The pain often awakens him at 1 or 2 A.M. An endoscopic examination and x-ray studies with barium showed normal stomach function but the presence of duodenal ulcers. Gastric analysis demonstrated that the gastric juice pH fell to 1.9 with pentagastrin stimulation (6 ug/kg s.c.). Fasting serum gastrin levels were normal. Recommended treatments for this individual follow.

1.  **Antacids:** 15–30 ml of liquid or two to four tablets one to three hours after each meal and at bedtime for six weeks.

2.  **Histamine (H₂) receptor antagonists:** cimetidine (300 mg) or ranitidine (150 mg) with each meal and at bedtime for four to eight weeks.

# Gastrointestinal Case Histories

## CASE 17

**QUESTIONS:**

1. What is the diagnosis of this individual?

2. What is the significance of doing a pentagastrin stimulation test?

3. What is the significance of doing a fasting serum gastrin level?

4. What is the function of taking antacids?

    *Counteracting acidity*

5. How do the H$_2$ antagonists act in treating this disorder?

6. What type of dietary and behavioral recommendations would also be suggested for this person?

Diet

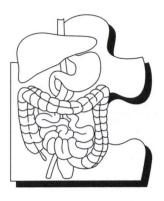

# Gastrointestinal Case Histories

## C A S E  18

A college student complained of pain, nausea with vomiting, and tenderness in the right lower quadrant. The pain was first vague and diffuse, then became more severe in the midepigastric region before localizing in the right lower quadrant. The pain was accentuated by movement, deep respiration, coughing, or sneezing. A mild fever of 102.2° F and a moderate leukocytosis (11,500/cu mm) were present. A marked tenderness was noted over the right lower quadrant at McBurney's point (one-third the distance between the anterior superior iliac spine and the umbilicus). The psoas sign (pain on passive hyperextension of the thigh) was also present.

# Gastrointestinal Case Histories

## C A S E  18

### QUESTIONS:

1. What is the diagnosis of this individual?

2. Describe in anatomical terms the location of the organ involved.

3. Locate the midepigastric region and McBurney's point on yourself. Draw an outline of the torso indicating the midepigastric region and McBurney's point.

4. What is the cause of the fever, pain, and leukocytosis in this person?

5. What is the usual treatment for this disorder?

# FOR YOUR CONSIDERATION: GASTROINTESTINAL

How familiar are you with some common categories of gastrointestinal drugs or agents? Match the drug category on the left with the indication or disorder on the right.

| | | |
|---|---|---|
| 1. Antacids | _____ | Gastric acids |
| 2. Anticholinergics | _____ | Gastric bloating |
| 3. Antidiarrheals | _____ | Diarrhea |
| 4. Antiemetics | _____ | Constipation |
| 5. Antiflatulents | _____ | Induces vomiting |
| 6. Antilipemics | _____ | Nausea, vomiting, dizziness |
| 7. Emetics | _____ | Adjunctive treatment for peptic ulcers and other GI disorders |
| 8. Histamine$_2$-receptor antagonists | | |
| | _____ | Duodenal and gastric ulcer treatment (short term) |
| 9. Laxatives | | |
| | _____ | Primary hyperlipidemia |

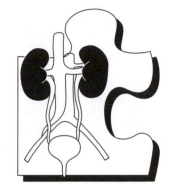

# Renal Case Histories

## C A S E  19

A 23-year-old male felt "puffy," weak, and tired for several months. He suddenly noticed his urine had a red to brown discoloration and the volume was minimal. He went to the emergency room of a nearby hospital and the following data was obtained upon examination and testing:

|  |  |  |  |
|---|---|---|---|
| **Hematology:** | Serum sodium | 125 mEq/L | *136-145* |
|  | Serum potassium | 6 mEq/L |  |
|  | Serum creatinine | 2.6 mg/dL |  |
|  | BUN | 10.0 mg/dL |  |
|  | pH (arterial) | 7.32 |  |
|  | Hematocrit | 25% |  |
| **Urinalysis:** | Appearance | Red to brown |  |
|  | Specific gravity | 1.025 |  |
|  | Blood | Positive |  |
|  | Glucose | Negative |  |
|  | Protein | Mild |  |
| **Renal Function Tests:** | GFR (glomerular filtration rate) | 40 ml/min | *120-125* |
|  | RBF (renal blood flow) | 280 ml/min | *1100-1200* |

# Renal Case Histories

## C A S E  19

### QUESTIONS:

1. What is the disorder of this individual? What situation(s) predispose an individual to this disorder? *Renal insufficiency — insulin glomerulo nephritis*

2. Define hyponatremia and hyperkalemia.

3. What is the cause of the hyponatremia and hyperkalemia?

4. Why is there blood in the urine?

   *infection WBC leak a RBC nephrosis leak*

5. How do the renal function tests for this individual compare to normal?

*pg 533*

6. What caused the "puffy" feeling?

7. What type of treatment does this person need?

8. Is this person a candidate for kidney dialysis? Explain your answer.

# Renal Case Histories

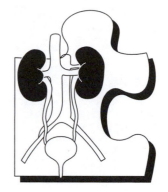

## C A S E  20

A 45-year-old construction worker complained of excruciating intermittent pain in the kidney area, radiating across the abdomen and into the genital area. He also had chills, fever, and nausea. He noticed increased frequency of urination and moderate hematuria. Pertinent 24-hour urinalysis findings indicated crystalline substances in the sediment identified as calcium in nature and a urinary calcium of 300 mg/day. X-ray findings indicated localized stones in the renal pelvis. This person was encouraged to increase his water intake to at least 1.0–1.5 L/day and slightly decrease his dietary calcium.

# Renal Case Histories

## C A S E  20

### QUESTIONS:

1.  This "calcium stone former" may have the hereditary condition known as idiopathic hypercalciuria. What does this mean?

2.  How do "stones," or calculi, form?

3.  Are calculi formed from minerals or compounds other than calcium? If so, give examples.

4.  List some ways renal calculi are removed.

# Renal Case Histories

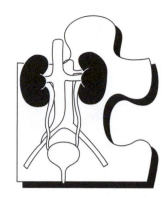

## CASE 21

*dehydration*   *Ketones in breakdown in fat*

A 26-year-old male with diabetes mellitus has developed renal failure. While waiting for a kidney transplant, he is on maintenance hemodialysis eight to ten hours three times each week. He is on a diet restricted in sodium (500 mg/day), potassium (2.6 gm/day), and protein as well as his usual diabetic diet. He has a shunt in his right wrist to allow for easy hookup to the dialysis machine. Prior to hemodialysis, his representative blood values are the following:

| | |
|---|---|
| Serum sodium | 120 mEq/L  136-145 |
| Serum potassium | 6.4 mEq/L  3.6 - 5.1 |
| Serum chloride | 102 mEq/L  98-106 |
| Serum creatinine | 16 mg/dL  0.7 - 1.3 |
| Hematocrit | 24%  40-54 |

The dialysis fluid in the kidney dialysis machine contains the following:

| | |
|---|---|
| Sodium | 134 mEq/L |
| Potassium | 2.6 mEq/L |
| Calcium | 2.5 mEq/L |
| Magnesium | 1.5 mEq/L |
| Chloride | 104 mEq/L |
| Sodium acetate | 36.6 mEq/L |
| Anhydrous dextrose | 2 gm/L |

# Renal Case Histories

## CASE 21

**QUESTIONS:**

1. What is hemodialysis?

2. Following eight to ten hours of hemodialysis, do you think the following blood values would be increased, decreased, or remain the same?

   Serum sodium: _____

   Serum potassium: _____

   Serum chloride: _____

   Serum creatinine: _____

   Hematocrit: _____

3. Why does anemia usually develop with maintenance dialysis?

4.  Why is hemodialysis required every two to three days for eight to ten hours/day for individuals with complete renal failure? (Flow rate of blood through the dialyzer is 150–300 ml/min.)




5.  Differentiate between hemodialysis and peritoneal dialysis.

# FOR YOUR CONSIDERATION: RENAL

How familiar are you with some common categories of renal drugs or agents? Match the drug category on the left with the indication or disorder on the right.

| | |
|---|---|
| 1. Acidifiers | _____ Hyperkalemia |
| 2. Alkalinizers | _____ Hypocalcemia |
| 3. Calcium chloride electrolyte replacement | _____ Hypokalemia |
| 4. Magnesium sulfate electrolyte replacement | _____ Hypomagnesemia |
| 5. Potassium chloride electrolyte replacement | _____ Metabolic acidosis |
| 6. Potassium-removing resins | _____ Metabolic alkalosis |
| 7. Urinary tract antiseptics | _____ Urinary tract infection |

# Endocrine Case Histories

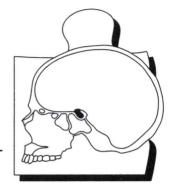

## C A S E  22

A 28-year-old male complained of abrupt polydipsia and polyuria. Blood and urine analyses provided the following results:

| | |
|---|---|
| Fasting blood glucose | 93 mg/dL |
| Serum sodium | 145 mEq/L |
| Serum potassium | 2.8 mEq/L |
| Urine specific gravity | <1.005 |
| Urine osmolality | <200 mOsm/L |
| Urine volume | 15 L/day |
| Urine glucose | 0 |

Water deprivation and hypertonic saline infusion do not cause a significant reduction in the polyuria and concentration of the urine. Complete water-deprivation results in the following:

| | |
|---|---|
| Urine specific gravity | 1.009 |
| Urine osmolality | 225 mOsm/L |

However, there is a significant concentration of the urine and a decrease in urinary output following administration of ADH.

# Endocrine Case Histories

## C A S E  22

**QUESTIONS:**

1. Define polydipsia and polyuria.

2. Why did the water deprivation and hypertonic saline infusion not result in a concentrated urine?

3. Describe the location of the disorder in this individual.

4.  A nasal spray containing a synthetic substance was self-administered to treat this condition. What type of compound was present in the nasal spray?

5.  Discuss the hypothalamic-pituitary-target organ pathway for this individual and indicate the normal and pathophysiological conditions involved.

# Endocrine Case Histories

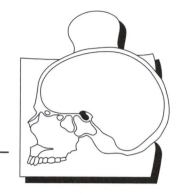

## C A S E  23

A 30-year-old female demonstrated a subtle onset of the following symptoms: dull facial expression; droopy eyelids; puffiness of the face and periorbital swelling; sparse, dry hair; dry, scaly skin; evidence of intellectual impairment; lethargy; a change of personality; bradycardia (60 b/min); a blood pressure of 90/70; anemia (hematocrit = 27); enlarged heart (upon radiological exam); constipation, and hypothermia. Plasma concentrations of total and free $T_4$ and $T_3$ follow:

|       | $T_4$       | $T_3$         |
|-------|-------------|---------------|
| Total | 3.0 ug/dL   | 0.14 ng/dL    |
| Free  | 0.6 ng/dL   | 0.01 ng/dL    |

Radioimmunoassay (RIA) of peripheral blood indicated elevated TSH levels. A TSH stimulation test did not increase the output of thyroid hormones from the thyroid gland.

# Endocrine Case Histories

## C A S E  23

### QUESTIONS:

1. What endocrine organ is involved here?

2. Is this a primary or secondary disorder? Why? (Is a TSH or TRH determination necessary for your diagnosis?)

3. Describe the feedback loop involved, indicating if there is an increased or decreased TSH level.

4. List several defects that could cause these symptoms.

5. Would you expect to find a palpable goiter? Explain your answer.

6. Describe a suitable treatment for this individual.

# Endocrine Case Histories

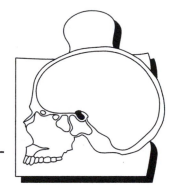

## C A S E  24

A 45-year-old male from the Midwest presented with the following symptoms during February: weakness, fatigue, orthostatic hypotension, weight loss, dehydration, and decreased cold tolerance. His blood chemistry values follow:

| | |
|---|---|
| Serum sodium | 128 mEq/L |
| Serum potassium | 6.3 mEq/L |
| Fasting blood glucose | 65 mg/dL |
| BUN | 4.5 mg/dL |
| Serum creatinine | 0.5 mg/dL |

Hematology tests resulted in the following values:

| | |
|---|---|
| Hematocrit | 50% |
| Leukocytes | 5000/cu mm |

He also noticed increased pigmentation (tanning) of both exposed and nonexposed portions of the body and back. A plasma cortisol determination indicated a low cortisol level. Following administration of ACTH, plasma cortisol did not rise significantly after sixty and ninety minutes. Endogenous circulating levels of ACTH were later determined to be significantly elevated.

# Endocrine Case Histories

PSS29
290

## CASE 24

**QUESTIONS:**

1. What endocrine organ is the site of the malfunction? Is this a primary or secondary disturbance?

   ADrenal cortex

2. What is the name of this disorder?  ADDISON's

3. Discuss the electrolyte disturbances resulting from this disorder.

   SODIUM IS Low

4. Discuss the metabolic disturbances resulting from this disorder.

5. What is the cause of the tanning? – Can't make anymore aldosterone

6. What type of replacement therapy would be required for this individual?




7. Diagram the feedback loop for this endocrine disorder.

# Endocrine Case Histories

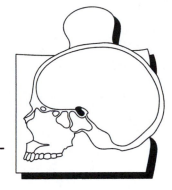

## C A S E  25

A 21-year-old noncompliant male with a history of type I (insulin-dependent) diabetes mellitus was found in a coma. His blood glucose was high, as well as his urine glucose, urine ketones, and serum ketones. His serum bicarbonate was <12 mEq/L. His respiration was exaggerated and his breath had an acetone odor. His blood pressure was 96/60 and his pulse weak and rapid (120).

# Endocrine Case Histories

## C A S E  25

**QUESTIONS:**

1. Define "noncompliant."

2. Is this person experiencing ketoacidosis or insulin shock? Explain your answer.

3. Why is the serum bicarbonate low?

4. What is the acid-base status of this individual?

5.  What is the cause of the dyspnea, hypotension, and tachycardia?




6.  What type of treatment does this person need?

# Endocrine Case Histories

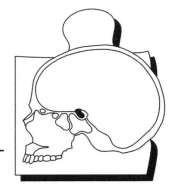

## C A S E  26

A 50-year-old male had a total thyroidectomy followed by thyroid hormone-replacement therapy. Thirty-six hours later he developed laryngeal spasms, a mild tetany, and cramps in the muscles of the hands and arms. The following tests were performed:

| | |
|---|---|
| Urine calcium | 20 mg/dL |
| Urine phosphorus | 0.1 g/day |
| Plasma calcium | 7.0 mg/dL |
| Plasma phosphorus | 5.0 mg/dL |

Calcium gluconate and vitamin D (calcitriol) were given orally each day and the tetany and laryngeal spasms were alleviated.

# Endocrine Case Histories

## C A S E  26

### QUESTIONS:

1. What endocrine disorder is present in this person?

2. What is the purpose of vitamin D administration with the calcium?

3. What caused the tetany and laryngeal spasms?

4. How is blood calcium normally maintained at its physiological level?

## FOR YOUR CONSIDERATION: ENDOCRINE

How familiar are you with some common categories of endocrine drugs or agents? Match the drug category on the left with the indication or disorder on the right.

1. Antidiabetic agents

2. Calcium regulating agents

3. Corticosteroids

4. Pituitary hormones

5. Thyroid hormones

6. Thyroid hormone antagonists

_____ Adrenal insufficiency; severe inflammation

_____ Cretinism

_____ Hyperglycemia

_____ Hyperthyroidism

_____ Pituitary diabetes insipidus; diagnostic test of adrenocortical function

_____ Postmenopausal osteoporosis; hypocalcemia

# Reproductive Case Histories

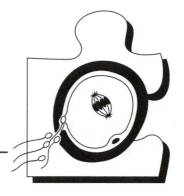

## CASE 27

Parents were concerned about their 16-year-old son for the following reasons: he had no deepening of his voice, scanty pubic and axillary hair growth, absence of beard and mustache growth, small penis, poor muscular development, and psychosocial immaturity.

Clinical evaluation indicated the following:

| | |
|---|---|
| Serum testosterone | 100 ng/dL |
| Sperm count | 10 million/ml semen |

The following tests were performed:

Clomiphene (a nonsteroidal, weak estrogen agonist that stimulates the release of gonadotropins) 100 mg/day for seven days: 0% increase in LH (50% is normal)

Gn-RH (100 ug I.V.): 0% increase in LH in twenty minutes (300% is normal)

HCG (5000 I.U., I.V.): 50% increase in plasma testosterone one to three days after injection

This person was subsequently treated with FSH at 25–75 U three times/week and HCG as described above. Sperm count and testosterone levels were both near normal after two months of treatment, and primary and secondary sex characteristics appeared.

# Reproductive Case Histories

## C A S E   27

### QUESTIONS:

1.  What is the endocrine disorder in this individual?

2.  Is this a primary or secondary disorder? Why?

3.  Why is HCG used in the treatment?

4.  Why would both FSH and HCG be needed in the treatment?

# Reproductive Case Histories

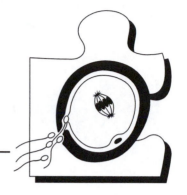

## C A S E   28

A 26-year-old female complained of severe, dull, aching pain, and cramping in the lower abdomen. There were no other physical findings. A laparoscopy revealed the presence of ectopic endometrial tissue on the uterine wall and ovaries. Danazol (a synthetic androgen and inhibitor of gonadotropins), 600 mg/day, was prescribed for up to nine months to inhibit ovulation, suppress the growth of the abnormal endometrial tissue, and achieve appreciable symptomatic relief, with a 30% possibility of conception after withdrawal of the therapy.

# Reproductive Case Histories

## C A S E  28

### QUESTIONS:

1. What is this condition called? What causes it?

2. What is ectopic endometrial tissue?

3. What is the rationale for using danazol, a gonadotropin inhibitor?

4. Why do you think oral contraceptives could also be used as a treatment?

# Reproductive Case Histories

## C A S E   29

A 25-year-old woman stated that it had been six weeks since her last menses. Her pregnancy test result was positive. By the sixth month of pregnancy, she felt irregular contractions of the uterus but no complications were present. After nine months, a healthy, 7 lb, 3 oz girl was delivered with no complications. Breast feeding was planned.

# Reproductive Case Histories

## CASE 29

**QUESTIONS:**

1.  What hormonal component is the basis of pregnancy tests?

2.  Following the positive pregnancy test, what hormonal mechanism prevented spontaneous abortion of the implanted embryo?

3.  What prevented the uterus from initiating labor before the designated delivery time?

4.  Describe the positive feedback systems that occur during labor and delivery.

5. What maintains milk production after birth?

6. Is it possible to get pregnant during the time of breast feeding? Explain your answer.

# FOR YOUR CONSIDERATION: REPRODUCTION

How familiar are you with some common categories of reproductive drugs or agents? Match the drug category on the left with the indication or disorder on the right.

| | |
|---|---|
| 1. Estrogens | _____ Anovulation, infertility |
| 2. Fertility agents | _____ Menopausal symptoms |
| 3. Gonadotropins | _____ Menstrual disorders; amenorrhea; abnormal uterine bleeding |
| 4. Progestins | |
| | _____ Ovulation inducers |

# Appendix A
## Medical Abbreviations

| | | | |
|---|---|---|---|
| < | lesser than | LH | luteinizing hormone |
| > | greater than | mEq | milliequivalents |
| ACTH | adrenocorticotropic hormone | mg | milligram |
| A-V | atrioventricular | ml | milliliter |
| B-blocker | beta-blocker | mm Hg | millimeters of mercury |
| BP | blood pressure | mmol | millimole |
| BUN | blood urea nitrogen | mOsm | milliosmole |
| CBC | complete blood count | ng | nanogram; $10^{-9}$g |
| CO | cardiac output | $PaCO_2$ | partial pressure of carbon dioxide in arterial blood |
| cu mm | cubic millimeter | | |
| dL | deciliter | PAH | para-aminohippuric acid |
| $FEV_{1.0}$ | forced expiratory volume at 1 second | $PaO_2$ | partial pressure of oxygen in arterial blood |
| FRC | functional residual capacity | pg | picagrams; $10^{-12}$g |
| FSH | follicle-stimulating hormone | RBF | renal blood flow |
| FVC | forced vital capacity | RIA | radioimmunoassay |
| g | gram | RPF | renal plasma flow |
| GFR | glomerular filtration rate | RVP | right ventricular pressure |
| Gn-RH | gonadotropin releasing hormone | S.C. | subcutaneous |
| HCG | human chorionic gonadotropin | $T_3$ | triiodothyronine |
| Hct | hematocrit | $T_4$ | thyroxine |
| Hgb | hemoglobin | TLC | total lung capacity |
| HR | heart rate | TPR | total peripheral resistance |
| I.U. | international unit | TRH | thyrotropin releasing hormone |
| I.V. | intravenous | TSH | thyroid-stimulating hormone |
| L | liter | U | unit |
| LAD | left anterior descending | ug | microgram; $10^{-6}$g |
| LAP | left atrial pressure | VC | vital capacity |
| LES | lower esophageal sphincter | | |

# Appendix B
## Reference Laboratory Values

| | | | |
|---|---|---|---|
| **Acid Phosphatase** | <3.0 ng/ml | **Chloride** | 98–106 mEq/L |
| **Albumin** (Adult) | 3.5–5.0 g/dl | **Cholesterol** | 140–250 mg/dL |
| **Alkaline Phosphatase** (Adult) | 20–70 U/L | **Complete Blood Count** (CBC) (Adult) | |
| **Ammonia** (Adult) | 11–35 um/L |   Leukocytes (WBC) | 4,800–10,800/ cu mm |
| **Amylase** | 25–125 U/L | | |
| **B₁₂** | 140–700 pg/ml |   Erythrocytes (RBC) | M: 4,600,000–6,200,000/ cu mm |
| **Bicarbonate** (Total $CO_2$) | 23–29 mmol/L | | F: 4,200,000–5,400,000/ cu mm |
| **Bilirubin** | | | |
|   Total | <0.2–1.0 mg/dL |   Hemoglobin | M: 14–18 g/dL |
|   Direct | <0.2 mg/dL | | F: 12–16 g/dL |
|   Indirect | <0.8 mg/dL |   Hematocrit | M: 40–54% |
| **Blood Pressure** | 120/80 mm Hg | | F: 37–47% |
| **Blood Urea Nitrogen** (BUN) | 7–18 mg/dL |   Thrombocytes (Platelets) | 150,000–450,000/ cu mm |
| **Calcium** | 4.2–5.3 mEq/L or 8.5–10.5 mg/dL |   *Differential* | |
| | |     Segmented neutrophils | 41–71% |
| **Cardiovascular System** | | | |
|   Cardiac Output (resting) | 3.5–5.5 L/min |     Bands | 5–10% |
|   *Pressures* | |     Lymphocytes | 24–44% |
|     Aorta | |     Monocytes | 3–7% |
|       Systole | 120 mm Hg |     Eosinophils | 1–3% |
|       Diastole | 80 mm Hg |     Basophils | 0–1% |
|     *Atrium (Mean)* | | **Cortisol** (8 AM) | 5.0–23.0 mg/dL |
|       Left | 2–12 mm Hg |     (4 PM) | 3.0–15.0 mg/dL |
|       Right | 0–5 mm Hg | **Creatine Phosphokinase** (CPK) | 25–145 Mu/mL |
|     *Pulmonary Artery* | | | |
|       Systole | 12–28 mm Hg | **Creatinine** | M: 0.7–1.3 mg/dL |
|       Diastole | 3–13 mm Hg | | F: 0.6–1.1 mg/dL |
|       Wedge | 3–11 mm Hg | **Fibrinogen** | 150–450 mg/dL |
|     *Ventricle* | | **Folic Acid** (Folate) | 2.0–21 ng/mL |
|       Left Systole | 120 mm Hg | **Glomerular Filtration Rate** (GFR) | 120–125 ml/min |
|       Left Diastole | 2–12 mm Hg | | |
|       Right Systole | 25 mm Hg | **Glucose** (Fasting) | 65–95 mg/dL |
|       Right Diastole | 0–5 mm Hg |     2 hours | <120 mg/dL |
|   Stroke Volume | 70 ml/beat |       postprandial | |

| | | | |
|---|---|---|---|
| **Heart Rate** | 70-80 beats/min | Inspiratory Capacity | M: 3.8 L |
| **Iron** | M: 65-175 ug/dL | (IC) | F: 2.4 L |
| | F: 50-170 ug/dL | Inspiratory Reserve Vol | M: 3.3 L |
| **Iron-Binding Capacity,** | 300-360 ug/dL | (IRV) | F: 2.0 L |
| Total (TIBC) | | Maximum Expiratory | >300 L/min |
| **Lactate Dehydrogenase** | 45-100 U/L | Flow Rate | |
| (LDH) | | Maximum Inspiratory | >400 L/min |
| **Lactic Acid** (Lactate) | 0.5-2.2 mmol/L | Flow Rate | |
| **Lipase** | 0-1.5 U/ml | Minute Ventilation = | 5-9 L/min |
| **Magnesium** | 1.6-2.6 mg/dL | Resp Rate × TV | |
| **Osmolality** (Serum) | 278-298 mOsm/L | $PaCO_2$ | 35-45 mm Hg |
| **Oxygen** (Arterial) | 75-100 mm Hg | $PaO_2$ | 80-100 mm Hg |
| **Oxygen Saturation** | | pH (Arterial) | 7.38-7.42 |
| Arterial | 94-100% | Residual Volume (RV) | M: 1.2 L |
| Venous | 60-85% | | F: 1.0 L |
| **Phosphorus** (Adult) | 2.7-4.5 mg/dL | Respiratory Rate | 10-16 breaths/min |
| **Potassium** | 3.6-5.1 mmol/L or | Tidal Volume (TV) | 0.5 L |
| | mEq/L | Total Lung Capacity = | M: 6.0 L |
| **Protein** | 6.0-7.8 g/dL | VC + RV | |
| **Prothrombin Time** (PT) | 11.5-13.5 seconds | | F: 4.2 L |
| **Pulmonary** | | Vital Capacity (VC) = | M: 4.8 L |
| Alveolar Ventilation | 4.0-5.0 L/min | TV + IRV + ERV | |
| Dead Space Volume | 150 ml | | F: 3.1 L |
| Expiratory Reserve Vol | | **Renal Blood Flow** (RBF) | 1100-1200 ml/min |
| (ERV) | | | (assuming a |
| | M: 1.0 L | | hematocrit of 45%) |
| | F: 0.7 L | **Sodium** | 136-145 mmol/L or |
| $FEV_{1.0}$/FVC | >75% | | mEq/L |
| Forced Expiratory Vol | 80% of VC | **Sperm Count** | 40-250 million/ml of |
| in 1 Second ($FEV_{1.0}$) | | | semen |
| Functional Residual | M: 2.2 L | **$T_3$ RIA** (Triiodothyronine) | 120-200 ng/dL |
| Capacity (FRC) | F: 1.8 L | **$T_3$ RU** (Resin uptake) | 26-35% |
| | | **$T_3$** (Triiodothyronine) **FREE** | 230-660 ng/dL |